Florian Ion
PETRESCU

A NEW ATOMIC MODEL

Germany
2012

Scientific reviewer:
Dr. Veturia CHIROIU

Honorific member of
Technical Sciences Academy of Romania
(ASTR)
PhD supervisor in Mechanical Engineering

Copyright

Title book: A New Atomic Model

Author book: Florian Ion Petrescu

© 2001-2012, Florian Ion PETRESCU

petrescuflorian@yahoo.com

ALL RIGHTS RESERVED. This book contains material protected under International and Federal Copyright Laws and Treaties. Any unauthorized reprint or use of this material is prohibited. No part of this book may be reproduced or transmitted in any form or by any means, electronic or mechanical, including photocopying, recording, or by any information storage and retrieval system without express written permission from the authors / publisher.

Manufactured and published by:
Books on Demand GmbH, Norderstedt
ISBN 978-3-8482-1894-3

Welcome! A Short Book Description

The movement of an electron around the atomic nucleus has today a great importance in many engineering fields. Electronics, aeronautics, micro and nanotechnology, electrical engineering, optics, lasers, nuclear power, computing, equipment and automation, telecommunications, genetic engineering, bioengineering, special processing, modern welding, robotics, energy and electromagnetic wave field is today only a few of the many applications of electronic engineering. This book presents shortly a new and original relation which calculates the radius with that the electron is running around the atomic nucleus.

This book presents, shortly, a new and original relation (20 & 20') who determines the radius with that, the electron is running around the nucleus of an atom.

In the picture number 1 one presents some electrons that are moving around the nucleus of an atom.

Scientific reviewer:

Dr. Veturia CHIROIU

Honorific member of
Technical Sciences Academy of Romania
(ASTR)
PhD supervisor in Mechanical Engineering

You are welcome to read the full book!

PRESENTATION

The movement of an electron around the atomic nucleus has today a great importance in many engineering fields.

Electronics, aeronautics, micro and nanotechnology, electrical engineering, optics, lasers, nuclear power, computing, equipment and automation, telecommunications, genetic engineering, bioengineering, special processing, modern welding, robotics, energy and electromagnetic wave field is today only a few of the many applications of electronic engineering.

This book presents, shortly, a new and original relation (20 & 20') who determines the radius with that, the electron is running around the nucleus of an atom.

In the picture number 1 one presents some electrons that are moving around the nucleus of an atom.

One utilizes, two times the Lorenz relation (5), the Niels Bohr generalized equation (7), and a mass relation (4) which it was deduced from the kinematics energy relation written in two modes: classical (1) and coulombian (2). Equalizing the mass relation (4) with Lorenz relation (5) one obtains the form (6) which is a relation between the squared electron speed (v^2) and the radius (r).

The second relation (8), between v^2 and r, it was obtained by equalizing the mass of Bohr equation (7) and the mass of Lorenz relation (5).

In the system (8) – (6) eliminating the squared electron speed (v^2), it determines the radius r, with that the electron is moving around the atomic nucleus; see the relation (20).

For a Bohr energetically level (n=a constant value), one determines now two energetically below levels, which form an electronic layer.

The author realizes by this a new atomic model, or a new quantum theory, which explains the existence of electron-clouds without spin [1-2].

Writing the kinematics energy relation in two modes, classical (1) and coulombian (2) one determines the relation (3).

From the relation (3), determining explicit the mass of the electron, it obtains the form (4) [2].

INTRODUCTION

This chapter presents, shortly, a new and original relation (20 & 20') who determines the radius with that, the electron is running around the nucleus of an atom [2].

In the picture number 1 one presents some electrons that are moving around the nucleus of an atom [1].

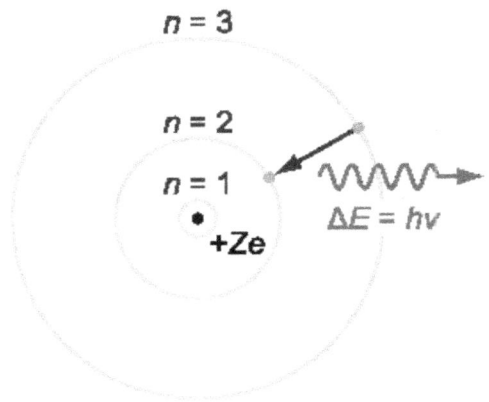

Fig. 1 *Electrons moving around the atomic nucleus;The atomic nucleus consists of nucleons (protons and neutrons)*

One utilizes, two times the Lorenz relation (5), the Niels Bohr generalized equation (7), and a mass relation (4) which it was deduced from the kinematics energy relation written in two modes: classical (1) and coulombian (2). Equalizing the mass relation (4) with Lorenz relation (5) one obtains the form (6) which is a relation between the squared electron speed (v^2) and the radius (r).

The second relation (8), between v^2 and r, it was obtained by equalizing the mass of Bohr equation (7) and the mass of Lorenz relation (5).

In the system (8) – (6) eliminating the squared electron speed (v^2), it determines the radius r, with that the electron is moving around the atomic nucleus; see the relation (20).

For a Bohr energetically level (n=a constant value), one determines now two energetically below levels, which form an electronic layer.

The author realizes by this a new atomic model, or a new quantum theory, which explains the existence of electron-clouds without spin [1-2].

Writing the kinematics energy relation in two modes, classical (1) and coulombian (2) one determines the relation (3).

From the relation (3), determining explicit the mass of the electron, it obtains the form (4) [2].

$$E_C = \frac{1}{2} m \cdot v^2 \qquad (1)$$

$$E_C = \frac{1}{8} \frac{Z \cdot e^2}{\pi \cdot \varepsilon_0 \cdot r} \qquad (2)$$

$$m \cdot v^2 = \frac{1}{4} \frac{Z \cdot e^2}{\pi \cdot \varepsilon_0 \cdot r} \qquad (3)$$

$$m = \frac{Z \cdot e^2}{4 \cdot \pi \cdot \varepsilon_0 \cdot v^2 \cdot r} \qquad (4)$$

Now, we write the known relation Lorenz (5), for the mass of a corpuscle in function of the corpuscle squared speed.

With the relations (4) and (5) ore obtains the first essential expression (6).

$$m = \frac{m_0 \cdot c}{\sqrt{c^2 - v^2}} \qquad (5)$$

$$\frac{m_0 \cdot c}{\sqrt{c^2 - v^2}} = \frac{Z \cdot e^2}{4 \cdot \pi \cdot \varepsilon_0 \cdot v^2 \cdot r} \qquad (6)$$

One utilizes now, the Niels Bohr generalized relation (7).

It uses for the second time the Lorenz relation (5) with the Bohr relation (7) and in this mode one obtains the second essential expression (8).

$$m = \frac{n^2 \cdot \varepsilon_0 \cdot h^2}{\pi \cdot r \cdot e^2 \cdot Z} \quad (7)$$

$$\frac{m_0 \cdot c}{\sqrt{c^2 - v^2}} = \frac{n^2 \cdot \varepsilon_0 \cdot h^2}{\pi \cdot r \cdot e^2 \cdot Z} \quad (8)$$

Now, one keeps just the two essential expressions (6 and 8). It writes (8) in the form (8').

$$\sqrt{c^2 - v^2} \cdot n^2 \cdot \varepsilon_0 \cdot h^2 = \pi \cdot r \cdot m_0 \cdot c \cdot e^2 \cdot Z \quad (8')$$

Elevating the relationship (8') to the square, to explicit the squared electron speed, it obtains the form (9).

$$v^2 = \frac{(n^4 \cdot \varepsilon_0^2 \cdot h^4 - \pi^2 \cdot r^2 \cdot m_0^2 \cdot e^4 \cdot Z^2) \cdot c^2}{n^4 \cdot \varepsilon_0^2 \cdot h^4} \quad (9)$$

The formula (9) can be put in the form (10), where the constant k takes the form (10').

$$v^2 = c^2 - k \cdot c^2 \cdot r^2 \quad (10)$$

$$k = \frac{\pi^2 \cdot m_0^2 \cdot e^4 \cdot Z^2}{n^4 \cdot \varepsilon_0^2 \cdot h^4} \quad (10')$$

Now one writes the essential relation (6) in the form (6').

$$4 \cdot m_0 \cdot c \cdot \pi \cdot \varepsilon_0 \cdot r \cdot v^2 = Z \cdot e^2 \cdot \sqrt{c^2 - v^2} \quad (6')$$

Then, putting the relation (6') at the square, it obtains the formula (6").

$$\begin{aligned} 16 \cdot m_0^2 \cdot c^2 \cdot \pi^2 \cdot \varepsilon_0^2 \cdot r^2 \cdot v^4 = \\ = Z^2 \cdot e^4 \cdot (c^2 - v^2) \end{aligned} \quad (6")$$

In the relation (6") one introduce the squared velocity of the electron, taken from the expression (10) and one obtains the formula (11).

$$\begin{aligned} 16 \cdot m_0^2 \cdot \pi^2 \cdot \varepsilon_0^2 \cdot (c^2 - k \cdot c^2 \cdot r^2)^2 = \\ = Z^2 \cdot e^4 \cdot k \end{aligned} \quad (11)$$

The (11) relationship can be arranged in the form (12).

$$(c^2 - k \cdot c^2 \cdot r^2)^2 = \frac{Z^2 \cdot e^4 \cdot k}{16 \cdot m_0^2 \cdot \pi^2 \cdot \varepsilon_0^2} \qquad (12)$$

One squares the relation (12) and it obtains the expression (13).

$$(c^2 - k \cdot c^2 \cdot r^2) = \pm \frac{Z \cdot e^2 \cdot \sqrt{k}}{4 \cdot m_0 \cdot \pi \cdot \varepsilon_0} \qquad (13)$$

The relation (13) can be arranged to the form (14).

$$k \cdot c^2 \cdot r^2 = c^2 \mp \frac{Z \cdot e^2 \cdot \sqrt{k}}{4 \cdot m_0 \cdot \pi \cdot \varepsilon_0} \qquad (14)$$

From relation (14) it explicit the squared electron radius and one obtains the relation (15).

$$r^2 = \frac{1}{k} \mp \frac{Z \cdot e^2}{4 \cdot m_0 \cdot \pi \cdot \varepsilon_0 \cdot \sqrt{k} \cdot c^2} \qquad (15)$$

Now, one exchange in the relation (15), the constant k with its expression (10') and it obtains the relation (16).

$$r^2 = \frac{n^4 \cdot \varepsilon_0^2 \cdot h^4}{\pi^2 \cdot m_0^2 \cdot e^4 \cdot Z^2} \mp \frac{n^2 \cdot h^2}{4 \cdot \pi^2 \cdot m_0^2 \cdot c^2} \qquad (16)$$

The expression (16) can be put in the form (17).

$$r^2 = \frac{n^4 \cdot \varepsilon_0^2 \cdot h^4}{\pi^2 \cdot m_0^2 \cdot e^4 \cdot Z^2} \cdot (1 \mp \frac{e^4 \cdot Z^2}{4 \cdot c^2 \cdot \varepsilon_0^2 \cdot h^2 \cdot n^2})$$ (17)

Extracting the square root of the expression (17), it obtains for the electron radius (r), the expression (18).

$$r = \pm \frac{n^2 \cdot \varepsilon_0 \cdot h^2}{\pi \cdot m_0 \cdot e^2 \cdot Z} \cdot \sqrt{1 \mp \frac{e^4 \cdot Z^2}{4 \cdot c^2 \cdot \varepsilon_0^2 \cdot h^2 \cdot n^2}}$$ (18)

Physically there is only the positive solution (19).

$$r = +\frac{n^2 \cdot \varepsilon_0 \cdot h^2}{\pi \cdot m_0 \cdot e^2 \cdot Z} \cdot \sqrt{1 \mp \frac{e^4 \cdot Z^2}{4 \cdot c^2 \cdot \varepsilon_0^2 \cdot h^2 \cdot n^2}} \quad (19)$$

The relation (19) is writing in final form (20) [3].

$$r = \frac{n^2 \cdot \varepsilon_0 \cdot h^2}{\pi \cdot m_0 \cdot e^2 \cdot Z} \cdot \sqrt{1 \mp \frac{e^4 \cdot ^2}{4 \cdot c^2 \cdot \varepsilon_0^2 \cdot h^2 \cdot n^2}} \quad (20)$$

The expression (20) it's not just a new theory for calculating the radius with that the electron is running around the nucleus of an atom, it is also a really new theory of an atomic model, or a new quantum theory.

For a value of the quantum number n (for a constant atomic number Z), we haven't just one energetically level (like in the Bohr model).

Now we can find two energetically below levels, which form an electronic layer, an electronic cloud. For example, for n=1, we have two sublevels (two below levels) [1-2].

USED NOTATIONS

The permissive constant (the permittivity): $\varepsilon_0 = 8.85418 \cdot 10^{-12} [\dfrac{C^2}{N \cdot m^2}]$;

The Planck constant:
$h = 6.626 \cdot 10^{-34} [J \cdot s]$;

The rest mass of electron:

$m_0 = 9.1091 \cdot 10^{-31} [kg]$;

The Pythagoras number:

$\pi = 3.141592654$;

The electrical elementary load:

$e = -1.6021 \cdot 10^{-19} [C]$;

The light speed in vacuum:

$c = 2.997925 \cdot 10^8 [\frac{m}{s}]$;

n=the principal quantum number (the Bohr quantum number);

Z=the number of protons from the atomic nucleus (the atomic number) [2].

DETERMINING THE TWO DIFFERENT ELECTRON SPEED VALUES

Relationship (6") may be written in the form (6"') [2].

$$16 \cdot m_0^2 \cdot c^2 \cdot \pi^2 \cdot \varepsilon_0^2 \cdot r^2 \cdot v^4 + Z^2 \cdot e^4 \cdot v^2 - Z^2 \cdot e^4 \cdot c^2 = 0 \qquad (6''')$$

It can see easily that the relation (6"') represents a two degree equation in v^2.

One calculates v^2 with the formula (6^{IVa}).

$$v_{1,2}^2 = \frac{-Z^2 \cdot e^4 \pm \sqrt{Z^4 \cdot e^8 + 8^2 \cdot m_0^2 \cdot \pi^2 \cdot \varepsilon_0^2 \cdot c^4 \cdot Z^2 \cdot e^4 \cdot r^2}}{2 \cdot 16 \cdot m_0^2 \cdot c^2 \cdot \pi^2 \cdot \varepsilon_0^2 \cdot r^2}$$
(6^{IVa})

Physically there is just the positive solution, and one keeps it for the relation (6^{IV}) (only the positive sign) [2].

$$v^2 = \frac{-Z^2 \cdot e^4 + \sqrt{Z^4 \cdot e^8 + 8^2 \cdot m_0^2 \cdot \pi^2 \cdot \varepsilon_0^2 \cdot c^4 \cdot Z^2 \cdot e^4 \cdot r^2}}{2 \cdot 16 \cdot m_0^2 \cdot c^2 \cdot \pi^2 \cdot \varepsilon_0^2 \cdot r^2}$$
(6^{IV})

It can thinks that the relation (6^{IV}) gives only one solution for the electron squared speed (v^2), but really there is two solutions for this parameter, v^2, because the value of the squared radius (r^2) gives two physically solutions. It put the relation (6^{IV}) in the form (6^V) [2].

$$v_{1,2}^2 = \frac{-1+\sqrt{1+\dfrac{8^2 \cdot m_0^2 \cdot \pi^2 \cdot \varepsilon_0^2 \cdot c^2}{Z^2 \cdot e^4} \cdot c^2 \cdot r^2}}{\dfrac{1}{2} \cdot \dfrac{8^2 \cdot m_0^2 \cdot c^2 \cdot \pi^2 \cdot \varepsilon_0^2}{Z^2 \cdot e^4} \cdot r^2} \qquad (6^V)$$

The formula (6^V) can be written in the form (6^{VI}), where the constant k_1 takes the form (6^{VII}) [2].

$$v_{1,2}^2 = \frac{\sqrt{1+k_1 \cdot c^2 \cdot r^2} - 1}{\dfrac{k_1}{2} \cdot r^2} \qquad (6^{VI})$$

$$k_1 = \frac{8^2 \cdot m_0^2 \cdot \pi^2 \cdot \varepsilon_0^2 \cdot c^2}{Z^2 \cdot e^4} \qquad (6^{VII})$$

Now one starts with relation (6^{VI}) who can be written in the form (21).

$$v^2 = \frac{2 \cdot c^2}{\sqrt{1 + k_1 \cdot c^2 \cdot r^2} + 1} \qquad (21)$$

One notes the radical with R (see the relation 22).

$$R = \sqrt{1 + k_1 \cdot c^2 \cdot r^2} \qquad (22)$$

In relation (22) one introduces for r^2 the expression (20) and it obtains the form (22').

$$R = \sqrt{1 + \frac{k_1 \cdot c^2}{k} \cdot (1 \mp \frac{2 \cdot \sqrt{k}}{c \cdot \sqrt{k_1}})} \qquad (22')$$

In relation (22') one exchanges the two constant k_1 and k with the two values from expressions (6^{VII}) respective (10') and it obtains for (22') the form (22") [2].

$$R = \sqrt{1 + \frac{8^2 m_0^2 \cdot \pi^2 \cdot \varepsilon_0^2 \cdot c^4 \cdot n^4 \cdot \varepsilon_0^2 \cdot h^4}{Z^2 \cdot e^4 \cdot \pi^2 \cdot m_0^2 \cdot e^4 \cdot Z^2} \cdot (1 \mp \frac{2\pi \cdot m_0 \cdot e^4 \cdot Z^2}{8n^2 \cdot \varepsilon_0^2 \cdot h^2 \cdot c^2})}$$
(22")

One put the expression (22") in the form (22''').

$$R = \sqrt{1 + \frac{8^2 \cdot \varepsilon_0^4 \cdot c^4 \cdot h^4 \cdot n^4}{e^8 \cdot Z^4}(1 \mp \frac{e^4 \cdot Z^2}{4\varepsilon_0^2 \cdot c^2 \cdot h^2 \cdot n^2})}$$

(22''')

The expression (22''') will be written in the form (22^{IV}).

$$R = \sqrt{1 + \frac{8^2 \cdot \varepsilon_0^4 \cdot c^4 \cdot h^4 \cdot n^4}{e^8 \cdot Z^4} \mp \frac{2 \cdot 8 \cdot \varepsilon_0^2 \cdot c^2 \cdot h^2 \cdot n^2}{e^4 \cdot Z^2}}$$

(22^{IV})

The expression (22^{IV}) can be restricted to the forms (22^V) and (22^{VI}).

$$R = \sqrt{\left(1 \mp \frac{8 \cdot \varepsilon_0^2 \cdot c^2 \cdot h^2 \cdot n^2}{e^4 \cdot Z^2}\right)^2}$$

(22^V)

$$R = \left|1 \mp \frac{8 \cdot \varepsilon_0^2 \cdot c^2 \cdot h^2 \cdot n^2}{e^4 \cdot Z^2}\right| \qquad (22^{VI})$$

One notes with E the expression (23).

$$E = \frac{8 \cdot \varepsilon_0^2 \cdot c^2 \cdot h^2}{e^4} \cdot \frac{n^2}{Z^2} \qquad (23)$$

This expression must be evaluated.

$$E = \frac{8 \cdot 8.85418^2 \cdot 10^{-24} \cdot 2.997925^2 \cdot 10^{16}}{1.6021^4 \cdot 10^{-76}} \cdot \frac{6.626^2 \cdot 10^{-68} \cdot n^2}{Z^2} = \frac{37564.06551 \cdot n^2}{Z^2} \qquad (23')$$

For Zmax=92, we have a minimum of expression E (23"):

$$E_{min} = 4.438098477 \cdot n^2 \qquad (23")$$

It can see easily that Emin > 1:

$$E_{min} \succ 1 \qquad (24)$$

Now, one can write the expression (22^{VI}) in the forms (22^{VII}) a, and b:

$$R_1 = E - 1 \qquad (22^{VIIa})$$

$$R_2 = E + 1 \qquad (22^{VIIb})$$

Only now the expression (21) can be evaluated and reduced to two forms (21[Ia]) and respective (21[Ib]):

$$v_1^2 = \frac{2 \cdot c^2}{E - 1 + 1} \qquad (21^{Ia})$$

$$v_2^2 = \frac{2 \cdot c^2}{E + 1 + 1} \qquad (21^{Ib})$$

The two relations take the forms (21[II]) a, and b:

$$v_1^2 = \frac{c^2}{\dfrac{E}{2}} \qquad (21^{IIa})$$

$$v_2^2 = \frac{c^2}{\dfrac{E}{2}+1} \qquad (21^{IIb})$$

If one replaces E with its expression (23) it obtains for the electron speeds the relations (21^{III}) a, and b [2].

$$v_1^2 = \frac{e^4 \cdot Z^2}{4 \cdot \varepsilon_0^2 \cdot h^2 \cdot n^2} \qquad (21^{IIIa})$$

$$v_2^2 = \frac{c^2}{\dfrac{4 \cdot \varepsilon_0^2 \cdot c^2 \cdot h^2 \cdot ^2}{e^4 \cdot Z^2}+1} \qquad (21^{IIIb})$$

DETERMINING THE MASSES AND THE ENERGY OF THE ATOMIC ELECTRON IN MOVEMENT

The exact squared speeds can be written in the forms (25, 26) [2].

$$r_- = r_1 \Rightarrow v_1^2 = \frac{e^4 \cdot Z^2 \cdot c^2}{4 \cdot \varepsilon_0^2 \cdot c^2 \cdot h^2 \cdot n^2} \qquad (25)$$

$$r_+ = r_2 \Rightarrow v_2^2 = \frac{e^4 \cdot Z^2 \cdot c^2}{4 \cdot \varepsilon_0^2 \cdot c^2 \cdot h^2 \cdot n^2 + e^4 \cdot Z^2} \qquad (26)$$

With these velocities one can write the two adequate masses (27), (28) [2].

$$r_- = r_1 \Rightarrow m_1 = \frac{m_0}{\sqrt{1 - \dfrac{e^4 \cdot Z^2}{4 \cdot \varepsilon_0^2 \cdot c^2 \cdot h^2 \cdot n^2}}} \qquad (27)$$

$$r_+ = r_2 \Rightarrow$$

$$m_2 = \frac{m_0}{\sqrt{1 - \dfrac{e^4 \cdot Z^2}{4 \cdot \varepsilon_0^2 \cdot c^2 \cdot h^2 \cdot n^2 + e^4 \cdot Z^2}}} \qquad (28)$$

The total electron energy can be written in the forms (29) and (30) [2].

$$r_- = r_1 \Rightarrow W_1 = \frac{m_0 \cdot c^2}{\sqrt{1 - \frac{e^4 \cdot Z^2}{4 \cdot \varepsilon_0^2 \cdot c^2 \cdot h^2 \cdot n^2}}} \qquad (29)$$

$$r_+ = r_2 \Rightarrow$$
$$W_2 = \frac{m_0 \cdot c^2}{\sqrt{1 - \frac{e^4 \cdot Z^2}{4 \cdot \varepsilon_0^2 \cdot c^2 \cdot h^2 \cdot n^2 + e^4 \cdot Z^2}}} \qquad (30)$$

The possible frequency of pumping, between the two near energetically below levels can be written in the form (31) [2].

$$\nu = \frac{W_1 - W_2}{h} = \frac{m_0 \cdot c^2}{h} \cdot$$

$$\cdot \left[\frac{1}{\sqrt{1 - \frac{e^4 \cdot Z^2}{4 \cdot \varepsilon_0^2 \cdot c^2 \cdot h^2 \cdot n^2}}} - \frac{1}{\sqrt{1 - \frac{e^4 \cdot Z^2}{4 \cdot \varepsilon_0^2 \cdot c^2 \cdot h^2 \cdot n^2 + e^4 \cdot Z^2}}} \right]$$

(31)

THE *POSSIBLE* LASER FREQUENCIES

In the table 1, one can see the possible LASER pumping frequencies (all in visible domain $4.34 \times 10^{14} \div 6.97 \times 10^{14}$ [Hz]), calculated for different principal quantum number n.

The possible L A S E R pumping frequencies   Table 1

n	Z	[zH]ν	Element	n	Z	[zH]ν	Element
2	15=5.54942E14	P			78=4.43344E+14	Pt	
	22=5.072E14	Ti			79=4.66537E+14	Au	
3	23=6.0598E14	V			80=4.90629E+14	Hg	
	29=4.8452E+14	Cu			81=5.15642E+14	Tl	
	30=5.54942E+14	Zn			82=5.41601E+14	Pb	
4	31=6.32782E+14	Ga			83=5.68529E+14	Bi	
	36=4.71283E+14	Kr			84=5.96449E+14	Po	
	37=5.25911E+14	Rb			85=6.25386E+14	At	
	38=5.8516E+14	Sr			86=6.55364E+14	Rn	
5	39=6.49284E+14	Y	11	87=6.86408E+14	Fr	
	43=4.6261E+14	Tc			85=4.41451E+14	At	
	44=5.072E+14	Ru			86=4.6261E+14	Rn	
	45=5.54942E+14	Rh			87=4.8452E+14	Fr	
	46=6.0598E+14	Pd			88=5.072E+14	Ra	
6	47=6.60463E+14	Ag			89=5.30668E+14	Ac	
	50=4.56488E+14	Sn			90=5.54942E+14	Th	
	51=4.94145E+14	Sb			91=5.8004E+14	Pa	
	52=5.34086E+14	Te			92=6.0598E+14	U	
	53=5.76403E+14	I			93=6.32782E+14	Np	
	54=6.21189E+14	Xe			94=6.60463E+14	Pu	
7	55=6.68536E+14	Cs	12	95=6.89044E+14	Am	

	57	=4.51937E+14	La		92	=4.39854E+14	U
	58	=4.8452E+14	Ce		93	=4.59306E+14	Np
	59	=5.18835E+14	Pr		94	=4.79396E+14	Pu
	60	=5.54942E+14	Nd		95	=5.00139E+14	Am
	61	=5.92904E+14	Pm		96	=5.21548E+14	Cm
	62	=6.32782E+14	Sm		97	=5.43638E+14	Bk
8	63	=6.7464E+14	Eu		98	=5.66422E+14	Cf
	64	=4.48422E+14	Gd		99	=5.89916E+14	Es
	65	=4.77132E+14	Tb		100	=6.14134E+14	Fm
	66	=5.072E+14	Dy		101	=6.39091E+14	Md
	67	=5.38669E+14	Ho		102	=6.64801E+14	No
	68	=5.71581E14	Er	13	103	=6.9128E+14	Lw
	69	=6.0598E+14	Tm		99	=4.38489E+14	Es
	70	=6.4191E+14	Yb		100	=4.56488E+14	Fm
9	71	=6.79416E+14	Lu		101	=4.75037E+14	Md
	71	=4.45624E+14	Lu		102	=4.94145E+14	No
	72	=4.71283E+14	Hf		103	=5.13824E+14	Lr
	73	=4.98035E+14	Ta		104	=5.34086E+14	Rf
	74	=5.25911E+14	W	14	105	=5.54942E+14	Db
	75	=5.54942E+14	Re				
	76	=5.8516E+14	Os				
	77	=6.16596E+14	Ir				
	78	=6.49284E+14	Pt				
10	79	=6.83255E+14	Au				

THE LASER FREQUENCIES AND CONCLUSIONS

If the second speed value does not exist physically, we must calculate the new atomic model just for the new first value, with the next relations:

$$r = \frac{n^2 \cdot \varepsilon_0 \cdot h^2}{\pi \cdot m_0 \cdot e^2 \cdot Z} \cdot \sqrt{1 - \frac{e^4 \cdot Z^2}{4 \cdot c^2 \cdot \varepsilon_0^2 \cdot h^2 \cdot n^2}} \quad (20')$$

$$v^2 = \frac{e^4 \cdot Z^2}{4 \cdot \varepsilon_0^2 \cdot h^2 \cdot n^2} \quad (25')$$

$$m = \frac{m_0}{\sqrt{1 - \frac{e^4 \cdot Z^2}{4 \cdot \varepsilon_0^2 \cdot c^2 \cdot h^2 \cdot n^2}}} \quad (27')$$

$$W = \frac{m_0 \cdot c^2}{\sqrt{1 - \frac{e^4 \cdot Z^2}{4 \cdot \varepsilon_0^2 \cdot c^2 \cdot h^2 \cdot n^2}}} \quad (29')$$

$$\gamma = \frac{m_0 \cdot c^2}{h} \left(\frac{1}{\sqrt{1 - \frac{e^4 \cdot Z^2}{4 \cdot \varepsilon_0^2 \cdot c^2 \cdot h^2 \cdot n_1^2}}} - \frac{1}{\sqrt{1 - \frac{e^4 \cdot Z^2}{4 \cdot \varepsilon_0^2 \cdot c^2 \cdot h^2 \cdot n_2^2}}} \right)$$
(31')

The pumping frequency required to achieve the transition of the electrons between two energetically levels can be written in the form (31').

In the table 2, one can see the LASER pumping frequencies.

All frequencies are outside visible area. One can make Ultraviolet Frequency-X ray LASER.

The bold value can be used to make a Rubin (Crystal) LASER.

The paper realizes a new atomic model and a new quantum theory (relation 20').

It determines as well the frequency of pumping for the transition between two energetically levels, with possible applications in LASER, MASER, IRASER industry (relation 31').

Table 2. The pumping frequencies, between two nearer level

Z	ν	El n₁-n₂	Z	ν	Element	Z	ν	Element
1		H	2		He	3	2.22122E+16	Li 1-2
4	3.95022E+16	Be 1-2	5	6.17499E+16	B 1-2	6	8.89688E+16	C 1-2
7	1.21175E+17	N 1-2	8	1.58388E+17	O 1-2	9	2.00631E+17	F 1-2
10	2.47929E+17	Ne 1-2	11	5.53738E+16	Na 2-3	12	6.59213E+16	Mg 2-3
13	7.73939E+16	Al 2-3	14	8.97936E+16	Si 2-3	15	1.03123E+17	P 2-3
16	1.17383E+17	S 2-3	17	1.32578E+17	Cl 2-3	18	1.48709E+17	Ar 2-3
19	5.7866E+16	K 3-4	20	6.41348E+16	Ca 3-4	21	7.07288E+16	Sc 3-4
22	7.76485E+16	Ti 3-4	23	8.48944E+16	V 3-4	24	**9,24672E+16**	Cr 3-4
25	1.00368E+17	Mn 3-4	26	1.08596E+17	Fe 3-4	27	1.17153E+17	Co 3-4
28	1.2604E+17	Ni 3-4	29	1.35258E+17	Cu 3-4	30	1.44806E+17	Zn 3-4
31	1.54686E+17	Ga 3-4	32	1.64899E+17	Ge 3-4	33	1.75446E+17	As 3-4
34	1.86327E+17	Se 3-4	35	1.97544E+17	Br 3-4	36	2.09097E+17	Kr 3-4
37	1.01887E+17	Rb 4-5	38	1.07502E+17	Sr 4-5	39	1.1327E+17	Y 4-5
40	1.19192E+17	Zr 4-5	41	1.25268E+17	Nb 4-5	42	1.31498E+17	Mo 4-5
43	1.37882E+17	Tc 4-5	44	1.44421E+17	Ru 4-5	45	1.51116E+17	Rh 4-5
46	1.57966E+17	Pd 4-5	47	1.64972E+17	Ag 4-5	48	1.72134E+17	Cd 4-5
49	1.79453E+17	In 4-5	50	1.86928E+17	Sn 4-5	51	1.94561E+17	Sb 4-5
52	2.02352E+17	Te 4-5	53	2.10301E+17	I 4-5	54	2.18408E+17	Xe 4-5
55	1.22612E+17	Cs 5-6	56	1.2715E+17	Ba 5-6	57	1.31772E+17	La 5-6
58	1.36479E+17	Ce 5-6	59	1.41271E+17	Pr 5-6	60	1.46147E+17	Nd 5-6
61	1.51109E+17	Pm 5-6	62	1.56157E+17	Sm 5-6	63	1.6129E+17	Eu 5-6
64	1.66508E+17	Gd 5-6	65	1.71813E+17	Tb 5-6	66	1.77203E+17	Dy 5-6
67	1.8268E+17	Ho 5-6	68	1.88243E+17	Er 5-6	69	1.93893E+17	Tm 5-6
70	1.9963E+17	Yb 5-6	71	2.05453E+17	Lu 5-6	72	2.11364E+17	Hf 5-6
73	2.17362E+17	Ta 5-6	74	2.23448E+17	W 5-6	75	2.29621E+17	Re 5-6
76	2.35883E+17	Os 5-6	77	2.42232E+17	Ir 5-6	78	2.4867E+17	Pt 5-6
79	2.55197E+17	Au 5-6	80	2.61813E+17	Hg 5-6	81	2.68517E+17	Tl 5-6
82	2.75311E+17	Pb 5-6	83	2.82195E+17	Bi 5-6	84	2.89168E+17	Po 5-6
85	2.96231E+17	At 5-6	86	3.03385E+17	Rn 5-6	87	1.8618E+17	Fr 6-7
88	1.90549E+17	Ra 6-7	89	1.94972E+17	Ac 6-7	90	1.99447E+17	Th 6-7
91	2.03976E+17	Pa 6-7	92	2.08557E+17	U 6-7	93	2.13193E+17	Np 6-7
94	2.17881E+17	Pu 6-7	95	2.22624E+17	Am 6-7	96	2.2742E+17	Cm 6-7
97	2.3227E+17	Bk 6-7	98	2.37174E+17	Cf 6-7	99	2.42131E+17	Es 6-7
100	2.47144E+17	Fm 6-7	101	2.5221E+17	Md 6-7	102	2.57331E+17	No 6-7
103	2.62506E+17	Lr 6-7	104	2.67736E+17	Rf 6-7	105	2.73021E+17	Db 6-7

THE RELATIONSHIPS

Determining the ray of an electron moving on an orbit around an atom

The main relationships 1 and 2 are written [2]:

$$\left.\begin{array}{l} \text{Kinetic energy } E_c = \frac{1}{2} \cdot m \cdot v^2 \\ \text{Coulomb form } E_C = \frac{1}{8} \cdot \frac{Z \cdot e^2}{\pi \cdot \varepsilon_0 \cdot r} \end{array}\right\} \Rightarrow m = \frac{Z \cdot e^2}{4 \cdot \pi \cdot \varepsilon_0 \cdot r \cdot v^2} \right\} \Rightarrow$$

$$\text{Lorentz relation } m = \frac{m_0 \cdot c}{\sqrt{c^2 - v^2}}$$

$$\Rightarrow \frac{m_0 \cdot c}{\sqrt{c^2 - v^2}} = \frac{Z \cdot e^2}{4 \cdot \pi \cdot \varepsilon_0 \cdot r \cdot v^2} \Rightarrow \begin{cases} l \cdot r \cdot c \cdot v^2 = \sqrt{c^2 - v^2} \\ \text{with } l = \frac{4 \cdot \pi \cdot m_0 \cdot \varepsilon_0}{Z \cdot e^2} \end{cases}$$

(1)

Niels Bohr relation $m = \dfrac{\varepsilon_0 \cdot h^2 \cdot n^2}{\pi \cdot e^2 \cdot Z \cdot r}$

Lorentz relation $m = \dfrac{m_0 \cdot c}{\sqrt{c^2 - v^2}}$

$\Rightarrow \dfrac{m_0 \cdot c}{\sqrt{c^2 - v^2}} = \dfrac{\varepsilon_0 \cdot h^2 \cdot n^2}{\pi \cdot e^2 \cdot Z \cdot r} \Rightarrow$

$\Rightarrow \dfrac{\pi \cdot m_0 \cdot e^2 \cdot Z}{\varepsilon_0 \cdot h^2 \cdot n^2} \cdot r \cdot c = \sqrt{c^2 - v^2} \Rightarrow \begin{cases} k \cdot r \cdot c = \sqrt{c^2 - v^2} \\ \text{with} \quad k = \dfrac{\pi \cdot m_0 \cdot e^2 \cdot Z}{\varepsilon_0 \cdot h^2 \cdot n^2} \end{cases}$

(2)

It put the relationship 2 at the square and we obtain the formula 3.

$$v^2 = c^2 - k^2 \cdot r^2 \cdot c^2 \qquad (3)$$

3 is inserted in the relationship 1 and we obtain the relations 4.

$$\begin{cases}
l \cdot r \cdot c \cdot (c^2 - k^2 \cdot r^2 \cdot c^2) = \sqrt{c^2 - c^2 + k^2 \cdot r^2 \cdot c^2} \Rightarrow \\
\Rightarrow l \cdot r \cdot c \cdot c^2 \cdot (1 - k^2 \cdot r^2) = \sqrt{k^2 \cdot r^2 \cdot c^2} \\
\\
l \cdot r \cdot c \cdot c^2 \cdot (1 - k^2 \cdot r^2) = \pm r \cdot c \cdot k \Rightarrow \\
\Rightarrow l \cdot c^2 \cdot (1 - k^2 \cdot r^2) = \pm k \Rightarrow 1 - k^2 \cdot r^2 = \pm \dfrac{k}{l \cdot c^2} \Rightarrow \\
\\
r^2 = \dfrac{1}{k^2} \cdot \left(1 \mp \dfrac{k}{l \cdot c^2}\right) \Rightarrow r = \pm \dfrac{1}{k} \cdot \sqrt{1 \mp \dfrac{k}{l \cdot c^2}} \Rightarrow \\
\Rightarrow r = \dfrac{1}{k} \cdot \sqrt{1 \mp \dfrac{k}{l \cdot c^2}} \Rightarrow \\
\\
r = \dfrac{\varepsilon_0 \cdot h^2 \cdot n^2}{\pi \cdot m_0 \cdot e^2 \cdot Z} \cdot \sqrt{1 \mp \dfrac{\pi \cdot m_0 \cdot e^2 \cdot Z \cdot e^2 \cdot Z}{n^2 \cdot \varepsilon_0 \cdot h^2 \cdot 4 \cdot \pi \cdot m_0 \cdot \varepsilon_0 \cdot c^2}} \Rightarrow \\
r = \dfrac{\varepsilon_0 \cdot h^2 \cdot n^2}{\pi \cdot m_0 \cdot e^2 \cdot Z} \cdot \sqrt{1 \mp \dfrac{e^4 \cdot Z^2}{4 \cdot \varepsilon_0^2 \cdot h^2 \cdot n^2 \cdot c^2}}
\end{cases}$$

(4)

The final form (in 4) determines the ray of an electron running on an orbit around an atom. We have two r values at a single principal quantum number, n. It obtains a new and doubled relationship [2].

Determining the velocities of an electron which is running around an atom

From relationship 1 it obtains the speed of an electron to the square. We determine relationships numbered with 5.

$$\begin{cases} v^2 = \dfrac{2 \cdot c^2}{1 + \sqrt{1 + 4 \cdot c^4 \cdot r^2 \cdot l^2}} \Rightarrow v^2 = \dfrac{2 \cdot c^2}{1 + R} \\ \text{with} \quad R = \sqrt{1 + 4 \cdot c^4 \cdot r^2 \cdot l^2} \end{cases}$$

$$R = \sqrt{1 + 4 \cdot c^4 \cdot r^2 \cdot l^2} = \sqrt{1 + \dfrac{4 \cdot c^4 \cdot l^2}{k^2} \mp 2 \cdot \dfrac{2 \cdot c^2 \cdot l}{k}} =$$

$$= \sqrt{\left(1 \mp 2 \cdot \dfrac{c^2 \cdot l}{k}\right)^2} = \left|1 \mp 2 \cdot \dfrac{c^2 \cdot l}{k}\right| =$$

$$= \begin{cases} \dfrac{2 \cdot c^2 \cdot l}{k} - 1 = \dfrac{8 \cdot \varepsilon_0^2 \cdot h^2 \cdot n^2 \cdot c^2}{e^4 \cdot Z^2} - 1 \\ \dfrac{2 \cdot c^2 \cdot l}{k} + 1 = \dfrac{8 \cdot \varepsilon_0^2 \cdot h^2 \cdot n^2 \cdot c^2}{e^4 \cdot Z^2} + 1 \end{cases}$$

$$\text{with} \quad E = \dfrac{2 \cdot c^2 \cdot l}{k} > 1$$

$$v_-^2 = \dfrac{2 \cdot c^2}{1 + \dfrac{8 \cdot \varepsilon_0^2 \cdot h^2 \cdot n^2 \cdot c^2}{e^4 \cdot Z^2} - 1} =$$

$$= \dfrac{2 \cdot c^2}{\dfrac{8 \cdot \varepsilon_0^2 \cdot h^2 \cdot n^2 \cdot c^2}{e^4 \cdot Z^2}} = \dfrac{c^2}{\dfrac{4 \cdot \varepsilon_0^2 \cdot h^2 \cdot n^2 \cdot c^2}{e^4 \cdot Z^2}} = \dfrac{k}{l}$$

$$v_+^2 = \dfrac{2 \cdot c^2}{1 + \dfrac{8 \cdot \varepsilon_0^2 \cdot h^2 \cdot n^2 \cdot c^2}{e^4 \cdot Z^2} + 1} = \dfrac{2 \cdot c^2}{\dfrac{8 \cdot \varepsilon_0^2 \cdot h^2 \cdot n^2 \cdot c^2}{e^4 \cdot Z^2} + 2} =$$

$$= \dfrac{c^2}{\dfrac{4 \cdot \varepsilon_0^2 \cdot h^2 \cdot n^2 \cdot c^2}{e^4 \cdot Z^2} + 1} = \dfrac{kc^2}{lc^2 + k}$$

(5)

Determining the mass of the electron in movement

When the speeds are known is simple to find quickly the masses values (forms 6).

$$m_- = \frac{m_0}{\sqrt{1 - \dfrac{1}{\dfrac{4 \cdot \varepsilon_0^2 \cdot h^2 \cdot n^2 \cdot c^2}{e^4 \cdot Z^2}}}}$$

$$m_+ = \frac{m_0}{\sqrt{1 - \dfrac{1}{\dfrac{4 \cdot \varepsilon_0^2 \cdot h^2 \cdot n^2 \cdot c^2}{e^4 \cdot Z^2} + 1}}} \quad (6)$$

Determining the energy of the electron in movement

To determine the energy of an electron in movement, it multiplies the mass of an electron with the squared speed of light (using the Einstein relation), (forms 7).

$$W_- = \frac{m_0 \cdot c^2}{\sqrt{1 - \dfrac{1}{\dfrac{4 \cdot \varepsilon_0^2 \cdot h^2 \cdot n^2 \cdot c^2}{e^4 \cdot Z^2}}}}$$

$$W_+ = \frac{m_0 \cdot c^2}{\sqrt{1 - \dfrac{1}{\dfrac{4 \cdot \varepsilon_0^2 \cdot h^2 \cdot n^2 \cdot c^2}{e^4 \cdot Z^2} + 1}}} \quad (7)$$

Determining the frequencies of pumping

Finally, we can write the frequency of pumping between the two energetic sub levels, adjacent (see the form 8).

$$v = \frac{W_1 - W_2}{h} = \frac{m_0 \cdot c^2}{h} \cdot$$

$$\cdot \left(\frac{1}{\sqrt{1 - \dfrac{1}{\dfrac{4 \cdot \varepsilon_0^2 \cdot h^2 \cdot c^2 \cdot n^2}{e^4 \cdot Z^2}}}} - \frac{1}{\sqrt{1 - \dfrac{1}{\dfrac{4 \cdot \varepsilon_0^2 \cdot h^2 \cdot c^2 \cdot n^2}{e^4 \cdot Z^2} + 1}}} \right)$$

(8)

Notes utilized (used notations) (forms 9)

The permissive constant (the permittivity): $\varepsilon_0 = 8.85418 \cdot 10^{-12}$ $[\frac{C^2}{N \cdot m^2}]$

The Planck constant: $h = 6.626 \cdot 10^{-34}$ $[J \cdot s]$

The rest mass of electron: $m_0 = 9.1091 \cdot 10^{-31}$ $[kg]$

The Pythagora's number: $\pi = 3.141592654$

The electrical elementary load: $e = -1.6021 \cdot 10^{-19}$ $[C]$

The light speed in vacuum: $c = 2.997925 \cdot 10^8 [\frac{m}{s}]$

n = the principal quantum number (the Bohr quantum number)

Z = the number of protons from the atomic nucleus (the atomic number)

(9)

Table 3: The LASER frequencies of pumping (n=2-5)

Z	ע[Hz]	Element
15	=5.54942E14	P
22	=5.072E14	Ti
23	=6.0598E14	V
29	=4.8452E+14	Cu
30	=5.54942E+14	Zn
31	=6.32782E+14	Ga
37	=5.25911E+14	Rb
38	=5.8516E+14	Sr
39	=6.49284E+14	Y

CONCLUSIONS

All frequencies, calculated in the table 1, are outside of the visible domain ($4.34*10^{14} \div 6.97*10^{14}$ [Hz]).

Only the atmospheric elements, N and O, are located near the visible frequencies when n=1.

The bold value can be used to make a Rubin (Crystal) LASER.

For n=2-5 there are nine values indicated to make a LASER in the visible domain (see the table 2).

The substance is structured in this mode, that, we can obtain more energy, if one can penetrate it deeply. In this mode, we can check and extract, small portions of energy, but the total obtained energy will be bigger.

The atomic electrons are coupled. The transition between the two coupled electrons can give us more energy, in small portions.

First, we can make a stronger "Electromagnetic Amplification by the Stimulated Emission of Radiation" LASER (MASER), by pumping the energy between two sub levels, adjacent.

This paper briefly describes how to determine the relationships by which it calculates the ray of an electron moving on an orbit around an atom.

Now, it's the time to correct the length of the r radius (see the Cap. 5. (4)→(12)).

CORRECTING THE LENGTH OF THE R RADIUS

The main expression (2) can be written in the form (10).

$$r = \frac{1}{k} \cdot \sqrt{1 - \frac{v^2}{c^2}} \qquad (10)$$

The velocities have the forms (11), known.

$$\begin{cases} v_{-}^2 = \dfrac{k \cdot c^2}{l \cdot c^2} = \dfrac{k}{l} \\ v_{+}^2 = \dfrac{k \cdot c^2}{l \cdot c^2 + k} \end{cases} \qquad (11)$$

With the relations (11) the expression (10) takes the forms (12).

$$\begin{cases} r_- = \dfrac{1}{k} \cdot \sqrt{1 - \dfrac{k}{l \cdot c^2}} = \\ \quad = \dfrac{\varepsilon_0 \cdot h^2 \cdot n^2}{\pi \cdot m_0 \cdot e^2 \cdot Z} \cdot \sqrt{1 - \dfrac{e^4 \cdot Z^2}{4 \cdot \varepsilon_0^2 \cdot h^2 \cdot n^2 \cdot c^2}} \\ \\ r_+ = \dfrac{1}{k} \cdot \sqrt{1 - \dfrac{k}{l \cdot c^2 + k}} = \\ \quad = \dfrac{\varepsilon_0 \cdot h^2 \cdot n^2}{\pi \cdot m_0 \cdot e^2 \cdot Z} \cdot \sqrt{1 - \dfrac{e^4 \cdot Z^2}{4 \cdot \varepsilon_0^2 \cdot h^2 \cdot n^2 \cdot c^2 + e^4 \cdot Z^2}} \end{cases}$$

(12)

The values imposed by relations 12 are probably the real physical values, because the main relations 1 and 2 are verified in the same time by the relationships 12.

The velocities, masses, energies and frequency of pumping have not changed (see a recap in cap. 6, relations 13-16).

RECAP

$$\begin{cases} r_- = \dfrac{1}{k} \cdot \sqrt{1 - \dfrac{k}{l \cdot c^2}} = \\ \qquad = \dfrac{\varepsilon_0 \cdot h^2 \cdot n^2}{\pi \cdot m_0 \cdot e^2 \cdot Z} \cdot \sqrt{1 - \dfrac{e^4 \cdot Z^2}{4 \cdot \varepsilon_0^2 \cdot h^2 \cdot n^2 \cdot c^2}} \\ \\ r_+ = \dfrac{1}{k} \cdot \sqrt{1 - \dfrac{k}{l \cdot c^2 + k}} = \\ \qquad = \dfrac{\varepsilon_0 \cdot h^2 \cdot n^2}{\pi \cdot m_0 \cdot e^2 \cdot Z} \cdot \sqrt{1 - \dfrac{e^4 \cdot Z^2}{4 \cdot \varepsilon_0^2 \cdot h^2 \cdot n^2 \cdot c^2 + e^4 \cdot Z^2}} \end{cases}$$

(13)

$$\begin{cases} v_-^2 = \dfrac{2 \cdot c^2}{1 + \dfrac{8 \cdot \varepsilon_0^2 \cdot h^2 \cdot n^2 \cdot c^2}{e^4 \cdot Z^2} - 1} = \\[2em] \quad = \dfrac{2 \cdot c^2}{\dfrac{8 \cdot \varepsilon_0^2 \cdot h^2 \cdot n^2 \cdot c^2}{e^4 \cdot Z^2}} = \\[2em] \quad = \dfrac{c^2}{\dfrac{4 \cdot \varepsilon_0^2 \cdot h^2 \cdot n^2 \cdot c^2}{e^4 \cdot Z^2}} = \dfrac{k}{l} \\[3em] v_+^2 = \dfrac{2 \cdot c^2}{1 + \dfrac{8 \cdot \varepsilon_0^2 \cdot h^2 \cdot n^2 \cdot c^2}{e^4 \cdot Z^2} + 1} = \\[2em] \quad = \dfrac{2 \cdot c^2}{\dfrac{8 \cdot \varepsilon_0^2 \cdot h^2 \cdot n^2 \cdot c^2}{e^4 \cdot Z^2} + 2} = \\[2em] \quad = \dfrac{c^2}{\dfrac{4 \cdot \varepsilon_0^2 \cdot h^2 \cdot n^2 \cdot c^2}{e^4 \cdot Z^2} + 1} = \dfrac{k \cdot c^2}{l \cdot c^2 + k} \end{cases}$$

(14)

$$m_{-} = \frac{m_0}{\sqrt{1 - \dfrac{1}{\dfrac{4 \cdot \varepsilon_0^2 \cdot h^2 \cdot n^2 \cdot c^2}{e^4 \cdot Z^2}}}}$$

$$m_{+} = \frac{m_0}{\sqrt{1 - \dfrac{1}{\dfrac{4 \cdot \varepsilon_0^2 \cdot h^2 \cdot n^2 \cdot c^2}{e^4 \cdot Z^2} + 1}}}$$

(15)

$$W_- = \frac{m_0 \cdot c^2}{\sqrt{1 - \dfrac{1}{\dfrac{4 \cdot \varepsilon_0^2 \cdot h^2 \cdot n^2 \cdot c^2}{e^4 \cdot Z^2}}}}$$

$$W_+ = \frac{m_0 \cdot c^2}{\sqrt{1 - \dfrac{1}{\dfrac{4 \cdot \varepsilon_0^2 \cdot h^2 \cdot n^2 \cdot c^2}{e^4 \cdot Z^2} + 1}}}$$

(16)

Bibliography

[1] David Halliday, Robert, R., - *Physics, Part II,* Edit. John Wiley & Sons, Inc. - New York, London, Sydney, 1966;
[2] Petrescu F.I., *The movement of an electron around the atomic nucleus,* in ICOME 2010, Craiova, 2010.

www.ingramcontent.com/pod-product-compliance
Lightning Source LLC
Chambersburg PA
CBHW071217240526
45470CB00018B/2067